Inventions and Discoveries

Agriculture

WORLD
BOOK

a Scott Fetzer company

Chicago

www.worldbookonline.com

World Book, Inc.
233 N. Michigan Avenue
Chicago, IL 60601
U.S.A.

For information about other World Book publications, visit our Web site at **http://www.worldbookonline.com** or call **1-800-WORLDBK (967-5325)**.
For information about sales to schools and libraries, call **1-800-975-3250 (United States), or 1-800-837-5365 (Canada)**.

Editorial:
Editor in Chief: Paul A. Kobasa
Project Managers: Cassie Mayer, Michael Noren
Editor: Jake Bumgardner
Content Development: Odyssey Books
Writer: Cheryl Reifsnyder
Researcher: Cheryl Graham
**Manager, Contracts & Compliance
  (Rights & Permissions):** Loranne K. Shields
Indexer: David Pofelski

Graphics and Design:
Associate Director: Sandra M. Dyrlund
Manager, Graphics and Design: Tom Evans
Coordinator, Design Development and Production:
  Brenda B. Tropinski
Senior Designer: Isaiah W. Sheppard, Jr.
Contributing Photographs Editor: Carol Parden
Senior Cartographer: John M. Rejba

Pre-Press and Manufacturing:
Director: Carma Fazio
Manufacturing Manager: Steven K. Hueppchen
Production/Technology Manager: Anne Fritzinger

Picture Acknowledgments:

Front Cover: © Comstock/SuperStock.
Back Cover: Réunion des Musées Nationaux/Art Resource.

© Agripicture Images/Alamy Images 31; © Richard G. Bingham II, Alamy Images 9; © Corbis Premium RF/Alamy Images 37, 38; © CSI Productions/Alamy Images 40; © David R. Frazier Photolibrary/Alamy Images 11, 13, 23, 33; © Jeff Greenberg, Alamy Images 35; © Robert Harding Picture Library/Alamy Images 10, 20; © Grant Heilman Photography/Alamy Images 36; © Tony Hertz/Alamy Images 27; © Art Kowalsky, Alamy Images 18; © Gunter Marx, Alamy Images 15; © Middle East/Alamy Images 17; © Jeff Morgan, Alamy Images 17; © North Wind Picture Archives/Alamy Images 28, 34; © Phototake, Inc./Alamy Images 42; © Paul Thompson Images/Alamy Images 5; AP Wide World Photos 41; © Reunion des Musess Nationaux, Art Resource 14; © Nick Saunders/Barbara Heller Photo Library, London/Art Resource 6; Private Collection, Bridgeman Art Library 24; © Buffalo & Erie County Historical Society 32; Reviewing Performance by Walter Haskell Hinton, 1936. From the Story of John Deere. John Deere Art Collection, Deere & Company, Moline, IL 29; © Getty Images 43; Granger Collection 16, 22, 23, 25, 26, 27, 28, 30, 31; Library of Congress 19; © Hannelie Coetzee, Masterfile 11; © Christina Handley, Masterfile 29 © Carol Michel 9; Shutterstock 7, 19, 26, 33, 44

All maps and illustrations are the exclusive property of World Book, Inc.

Library of Congress Cataloging-in-Publication Data

Agriculture.
    p. cm. – (Inventions and discoveries)
  Summary: "An exploration of the transformative impact of inventions and discoveries in agriculture. Features include fact boxes, sidebars, biographies, and a timeline, glossary, list of recommended reading and Web sites, and index"–Provided by publisher.
  Includes index.
  ISBN 978-0-7166-0385-6
  1. Agricultural innovations–Juvenile literature. 2. Agriculture–History–Juvenile literature. I. World Book, Inc. II. Series.
S494.5.I5A3262 2009
630–dc22
                    2008040642

Inventions and Discoveries
Set ISBN: 978-0-7166-0380-1
Printed in China
1 2 3 4 5 12 11 10 09

▶ Table of Contents

There is a glossary of terms on pages 45-46. Terms defined in the glossary are in type **that looks like this** on their first appearance on any spread (two facing pages).

Before the development of agriculture, people got all their food by gathering wild plants, hunting, and fishing.

What is an invention?

An invention is a new device, new product, or new way of doing something. Inventions change the way people live. Before the car was invented, some people rode horses to travel long distances. Before the light bulb was invented, people used candles and similar sources of light to see at night. Almost two million years ago, the creation of the spear and the bow and arrow helped people hunt better. Later, the invention of new farming methods allowed people to stay in one place instead of wandering around in search of food. Today, inventions continue to change the way we live.

What is agriculture?

Agriculture is farming, or the raising of crops and farm animals. It is the world's oldest **industry** (business). Agriculture began about 10,000 years ago, when prehistoric people discovered how to grow crops and how to tame and raise animals.

The development of early farming methods made food production easier, faster, and more reliable. It allowed people to settle in permanent villages and try other activities, such as trade (the exchange of goods) and the arts. By making food more readily available, agriculture made possible the development of **civilization.**

Over time, scientific advances have continued to improve agriculture. Farmers and, later, scientists developed better breeds of **livestock,** better **varieties** of crop plants, and more effective chemicals that help plants grow and that combat insects and diseases which can harm plants. People also invented farm equipment that made the planting and harvesting of crops easier. These advances have helped increase the amount of crops a farm can produce. They have also decreased the amount of time and work required for farming.

Agriculture continues to be the world's most important industry. Agriculture feeds people throughout the world. It also provides materials for clothing, shelter, medicine, fuels, paints, and various other goods.

Dairy farms, like this one in Vermont, produce milk for people all over the world.

▶ The Domestication of Plants and Animals

This drawing from 1565 shows Inca farmers growing corn in Peru.

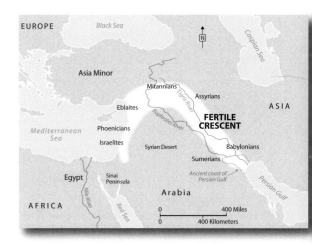

The waters of the Tigris and Euphrates rivers created the rich soil of the Fertile Crescent.

The first farmers lived in an area of the Middle East known as the Fertile Crescent. The Fertile Crescent began at the Mediterranean Sea, stretched between the Tigris and Euphrates rivers, and ended at the Persian Gulf. The region got its name because of its shape, which looks like a quarter-moon, or crescent. The region is also named for its **fertile** soil, which is rich with **nu-**trients (nourishing substances).

Farming began when tribes discovered that they could plant, raise, and **harvest** certain crops, such as barley and wheat. They also discovered how to keep certain animals, such as cattle, goats, and sheep. The **domestication** of plants and animals gave farmers much more control over their food sources. This, in turn, gave them more freedom in their lives.

At first, people did not depend entirely on farming for their food. They still used hunting and gathering methods. But with time, farming methods improved, and people could produce more and more food.

In some cases, agricultural methods spread when neighboring peoples came into contact with one another. In other cases, cultures developed agriculture independently.

Around 8000 B.C., people living in what are now the countries of Israel and Jordan were probably the first to depend chiefly on farming for food. Two thousand years later, people in northern Africa were herding cattle and growing grain.

Agriculture developed in Asia by around 5000 to 4000 B.C. Farmers there grew rice and **millet.** Between 4500 and 4000 B.C., farmers in southeastern Europe brought wheat and cattle to those living in northern Europe and Scandinavia.

In what is now Mexico, ancient people were growing maize (corn)

and beans by around 1500 B.C. By 1000 B.C., the people of eastern North America were growing gourds and sunflowers to supplement their hunting and gathering diets.

Farming continued to spread until it reached nearly every corner of the world.

The domestication of plants and animals gave ancient people more control over their food sources.

A CLOSER LOOK

The sprouting of a seed is a fascinating process. Seeds do not look alive. They do not move, grow, or do anything else on their own. But each seed is like an egg: it has an **embryo** inside it, the most basic form of all life. With water, oxygen, and warmth, the embryo springs to life and begins to grow both up and down. Once the first root and first green leaf have developed, the sprouting of the seed is complete.

The creation of tools like the hoe helped early farmers produce more food.

When prehistoric people planted the first seeds, they probably used simple planting methods, like scattering seeds on the ground. However, these planting methods would not have been very productive, as birds, mice, and other animals would likely eat the seeds.

People eventually learned to bury seeds in the ground to protect them and give them a chance to grow. They soon discovered that plants grew better in loose soil than in hard earth. So people began to break up the soil before planting by using sharp sticks, rocks, bones, and shells. These ob-jects were the forerunners of the hoe.

The hoe is a tool with a thin, flat blade used for **cultivating,** weeding, or loosening the earth around plants. The first hoes were jagged or sharp-ened rocks. Later, people attached wooden handles to the stones, creat-ing a hoe that could be used standing up rather than bent over on one's knees. Tools like the hoe allowed pre-historic people to do things they could not do with their bare hands.

With the development of metal tools, the hoe gained strength and sharpness, leading eventually to the scuffle hoe. This type of hoe is sharp-ened on both sides so that it can be either pushed forward or pulled back.

The hoe is useful for cutting weeds and roots from the soil and for breaking up hard dirt. Weeds compete with crops for sunlight, water, and **nutrients** found in the soil. Because of this, they reduce the amount and quality of the crops. Some types of weeds also attract harmful insects and plant diseases.

Such advances as modern farm machinery and **pesticides** (chemicals that kill pests) have replaced hoes on large farms in some countries. But in many parts of the world, including people's backyard gardens, the hoe continues to serve its timeless purpose.

Hoes come in all shapes and sizes, and each one is made for a special purpose. This young farmer (right) is clearing weeds with a garden hoe.

▶ The Plow

Ancient Egyptians used plows pulled by oxen, cattle, donkeys, and goats.

Throughout history, people have made improvements to tools and to the ways tools are used. As people became more experienced farmers, they invented the **plow,** a tool that made preparing the soil for planting a much faster and easier task.

The plow is a large tool used to turn over soil. Like a hoe, a plow loosens and breaks up hard soil. However, a plow can be used to turn over soil more quickly than a farmer could accomplish with just a hoe.

The earliest plows were called scratch plows because they scratched the earth. A scratch plow is a simple t-shaped frame that is made of wood. The frame holds a sharpened stick that is dragged through the **topsoil.**

Most scratch plows were pulled by an ox or two, as the force needed to drag a plow through the soil was greater than people could provide on their own. This simple plow cut a channel in the ground, where seeds could be planted and covered with loosened soil. These types of plows are still used in many countries.

Later improvements included the crooked plow, which was developed by people in ancient Greece. The crooked plow had an angled blade that was first made of bronze. Later, the blade was made of iron.

The plow made it easy for farmers to create a straight line of **tilled** soil, so ancient people began to plant their crops in rows. This farming technique is called **row cultivation.** Row cultivation simplified the planting process and made it easy to tell where crops were planted and where weeds were growing.

Today, most crops are still **cultivated** in rows. Farmers can prepare the soil, plant seeds, and apply **fertilizers** using modern farm equipment. Farmers can also kill weeds by plowing the ground between rows of crops.

Planting in rows makes it easier to **harvest** such oddly shaped crops as peanuts, corn, and sugar beets. Row

cultivation, together with improved farm machinery, has allowed fewer farmers to produce more food than at any other time in history.

Improvements to the plow continued over thousands of years. (See Steel Plow, pages 28-29.) Modern-day plows continue to be one of the most important pieces of farm equipment.

Water buffalo can help out on the farm, too. This one is pulling a plow in a rice field in the Philippines.

Today, farmers can use large plows pulled by tractors to plant crops in rows.

▶ Irrigation

Irrigation canals from the Nile River helped Egyptians raise crops and livestock.

To expand farming beyond **fertile** river valleys and **flood plains,** people had to find a way to bring water to areas that were dry. They did so through **irrigation**— that is, watering the land through artificial means for the growing of plants. Irrigation sounds like a simple idea, but it is quite difficult to bring water to dry land. Water is a heavy liquid and not easy to transport.

By 3000 B.C., people in Egypt built an elaborate system of canals that carried water to fields away from the Nile River. This expanded the amount of land that could support

crops. By this time, irrigation systems had also been built in China, India, Peru, and areas of the Middle East.

Irrigation methods and other agricultural advances allowed ancient farmers to produce more food than they needed, creating a **surplus** to be used in times of **drought,** extreme hunger, or war. This surplus freed people to do other things besides farm. They did not have to worry as much about where the next meal was coming from, so they took interest in other things. Cities grew and began to swell with traders (people who buy and sell goods), builders, craft workers, artists, and priests. Writing systems were improved and the sciences were further explored.

By A.D. 200, the people in ancient Rome had learned irrigation methods from farmers in the Middle East. **Roman engineers** built long canals and **aqueducts** to increase their own food production. An aqueduct is an artificial channel through which water is transported to the place where it is needed. The ancient Romans spread their knowledge of irrigation methods throughout Europe.

People in ancient Egypt developed an irrigation system that channeled water through canals, ditches, and **embankments** to the fields.

Today, many farmers irrigate the land using systems of pipes, drips, and sprinklers. Though the methods have changed, irrigation remains critical to farming in many parts of the world. In desert areas like Egypt, irrigation makes farming possible. In areas that receive rain only part of the year, such as California, irrigation allows farmers to grow crops all year long.

Sprinkler irrigation systems create miniature rainstorms for thirsty crops.

► Crop Rotation

The fields around this medieval castle show the rotation of different crops and livestock.

Such agricultural advances as the ox-drawn **plow** and **irrigation** systems made farming easier and more efficient. As a result, more land was used to grow crops, and more crops were produced. But over time, farmers noticed that the soil became less **fertile** when they repeatedly planted the same type of crop in a field.

Ancient **Roman** farmers began the practice of leaving half of every field **fallow** (unplanted) each year. The fallow soil could store **nutrients** and moisture for a crop the following year. Then, around the A.D. 200's, they developed a new planting system to help restore the soil's nutrients. This system is called **crop rotation.**

Crop rotation is one of the oldest methods of maintaining soil quality. Each year, a farmer switches the type of crop planted in a field. This replaces the nutrients that are removed from the soil by the crops that grew there the previous year.

For example, corn is a crop that absorbs nitrogen, a chemical element that is necessary for all plant growth. If a farmer planted corn in a field each year, the field would become infertile. However, a farmer can replace the nitrogen in the soil by planting such **legumes** as alfalfa, soybeans, and clover. Legumes are any of the plants that belong to the pea family.

These plants accumulate nitrogen in their roots.

In addition to replacing lost minerals and other substances, crop rotation provides numerous other benefits to the soil. Some crops are planted because they have deep roots, such as alfalfa and canola. Their root systems create underground pathways that help water penetrate the soil. Other rotation crops are planted to help keep soil in place on steep slopes to fight against **erosion.** Crop rotation also helps defend against insects and plant diseases and discourages the growth of weeds.

Today, many farmers have replaced crop rotation with the use of human-made chemicals to replenish the soil's nutrients or protect crops against disease and harmful insects. These chemicals allow farmers to plant the same crops each year.

Though chemical **fertilizers** and **pesticides** have greatly increased the world's food production, they have also contributed to environmental problems, such as water **pollution.** (See Fertilizers, pages 36-37; and Pesticides, pages 40-41.)

FUN FACT Soybeans are used not only for food products, but also for such goods as paint and cosmetics.

Today, many farmers still use crop rotation to keep soil productive.

▶ The Water Wheel and Water Mill

This illustration shows grain mills from 1200's France that were powered by the strength of the river current.

Human beings have long relied on grains as a source of food. Historians believe that people learned to make flour from grain sometime between 15,000 B.C. and 9000 B.C. They ground the grain between two large, flat stones. This turned the grain into a powdery flour or meal (coarsely ground grain) that could be used for cooking and baking.

Grinding grain by hand takes a great deal of time and energy, but ancient people eventually found an easier method. They connected **water wheels** to the grinding stones, creating the **water mill.** A **mill** is a machine that grinds grain.

Water wheels are circular machines that are powered by flowing or falling sources of water, like rivers and waterfalls. Water wheels and mills were developed independently in different parts of the world, but the technology for both was largely the same. As the wheel spins, it powers gears and grinding stones inside the mill.

The first water wheels had blades that dipped into the water. The strength of the water current provided the strength of the grinding

gears. Later, water wheels were powered by water that poured over the top of the wheel, filling a series of buckets. The weight of water filling the buckets turned the wheel. This type of water wheel can only generate limited amounts of power.

Most early water wheels were used primarily to make flour. They were also used in **irrigation** and metalworking, as well as in the process of making sugar, malt, and wool clothing, among other things. Water wheels remained a major source of power until the invention of the **steam engine** in the 1700's.

Today, advanced water-driven machines are still widely used to

Water wheels were used in El Fayoum, Egypt, nearly 2,000 years ago. Today, these irrigation wheels are driven by canals carrying water to the surrounding farmland.

generate electric power. Most modern water wheels are horizontal. They lie flat, like a merry-go-round, instead of up and down, like a Ferris wheel.

Modern water mills are built right over a river, so the wheel turns directly below the grinding gears.

Some companies continue to make flour using traditional water mills, such as this one in Wales, a part of the United Kingdom.

The Windmill

Between A.D. 500 and 900, people in the Middle East developed the **windmill,** a machine similar to the **water mill.** Early windmills were horizontal and had cloth sails like a sailboat. The wind pushed the sails in a circle like a merry-go-round. As with water mills, most of these early windmills were used to grind grain.

By the 1100's, windmills had spread to Europe and grown larger. Inventors found that windmills could produce more power with heavier sails that turned vertically like a pinwheel. The heavier sails fell faster, and their winglike design raised them back up just as quickly. This and other improvements eventually led to the development of the Dutch windmill.

Dutch windmills are heavy and bulky with four large sails made of cloth or wood. Some types of Dutch windmills are used to pump water out of low-lying land to make the land suitable for farming.

Windmills were important to the agricultural development of the United States and Canada during the 1800's and early 1900's. Farmers built their own small windmills to pump water to fields. This allowed crops to grow in areas that would have normally been too dry.

Today, windmills are still used all over the world. Some are used to pump water from underground wells, while others are used to grind

These "ground sailer" windmills in the Netherlands pump water near the New Meuse River.

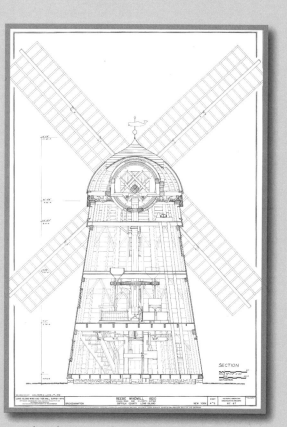

This diagram shows the complicated inner workings of the Beebe windmill in Bridgehampton, New York. It was built in 1820.

grain. But the majority of today's windmills are **wind turbines,** which are used to produce electric power. The blades of wind turbines are connected to a machine that can change the energy from wind into electric power. Large numbers of wind turbines harness the wind's power, providing a clean source of energy for modern **civilization.**

Thousands of wind turbines line the windy coasts and hills of countries all over the world.

▶ The Horse Harness

This farmer in Derbyshire, England, uses draft horses to pull farm equipment.

The use of oxen to pull **plows** began around 3000 B.C. in Egypt and other areas of the Middle East. Later, the ancient **Romans** introduced the ox-drawn plow to Europe, where its use continued for hundreds of years. But during the A.D. 900's, a new type of harness was introduced to Europe. This new harness made it possible to pull plows with horses as well as with oxen.

A harness is a device that attaches a horse, ox, or other animal to a piece of equipment, such as a plow or wagon. Ideally, it places the weight of the equipment on the animal's shoulders. This allows the animal to pull a heavier load.

Although horses had been tamed for riding by around 3000 B.C., the only available harnesses pressed against horses' windpipes. If a horse tried to pull a heavy load, the harness pressed into the animal's throat. This made it difficult or impossible for the horse to breathe.

The ox-drawn plow used a harness that fit only oxen, but the new harness allowed a horse to be hitched to the plow. A horse can pull

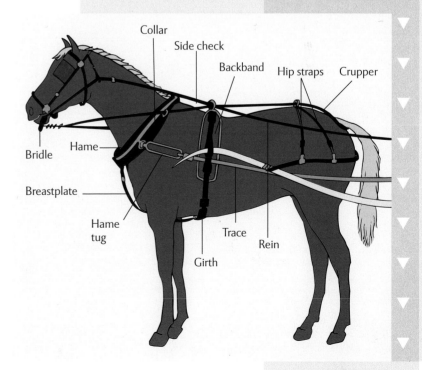

It was once common in many places to hook a dog up to a small cart or wagon to pull things around the farm. Small harnesses were made especially for dogs. In Holland, Belgium, and northern France, people used dogs to deliver milk around town!

mal's neck at the shoulders. The hames are two curved pieces attached to the sides of the collar. The traces are long straps, ropes, or chains that connect the harness to the equipment the animal will pull.

Animals like oxen, donkeys, mules, and horses are still used on farms in many parts of the world. However, most farmers in the United States, Canada, Australia, and the countries of western Europe have replaced horses with tractors and other modern farm equipment.

a plow three to four times faster than an ox, so the horse harness allowed farmers to **till** their land much more rapidly. After the introduction of the horse harness, horses gradually replaced oxen on many farms.

A harness is made up of several basic parts: the bridle, reins, collar, hames, and traces. The bridle is a set of straps that fit around the animal's head and connect to a bit that fits in its mouth. The bit allows a driver to control the animal's movements. The driver holds the reins—long straps connected to the bit—to direct the animal.

The collar fits around the ani-

A harness allows a horse to pull with its strong shoulders and legs.

Collar
Side check
Backband
Hip straps
Crupper
Bridle
Hame
Breastplate
Hame tug
Girth
Trace
Rein

The Seed Drill

In the early 1700's, farmers still sowed seeds by hand, scattering them on the ground. This method wasted much of the farmers' seed. It also resulted in an uneven **harvest,** as many seeds would never **germinate.** Farmers knew that they would get better results from planting single seeds by pushing them into the ground. However, planting seeds by hand took a very long time, so the scattering of seeds continued.

Sowing seeds got the job done, but modern methods allowed fewer seeds to produce more plants.

Inventors had long experimented with mechanical seed-planting devices. A Chinese version of such a device worked fairly well. Then in 1701, an English farmer named Jethro Tull built the first practical tool for planting seeds in rows. His invention was called the mechanical **seed drill.**

Tull got the idea for the seed drill while traveling through Europe to study new farming methods. While in Italy, he came across an early version of the seed drill. He brought home that technology and improved it.

Tull's seed drill made rows of small trenches (narrow ditches) in the soil, dropped seeds into them at just the right depth, and then covered them with soil. Pulled by a horse, the seed drill rolled on wheels and planted three rows at a time, using less seed than sowing by hand. This made farming much more efficient and productive.

Even with proven results, Tull's seed drill was not immediately successful in England. It quickly became popular with the American colonists, but it was not widely used until after

Tull's death late in the 1700's. Still, the mechanical seed drill was the first successful farm machine with moving parts. It is considered the ancestor of the modern farm machinery used today.

Tull's seed drill was one of the first in a series of inventions that led to the **Agricultural Revolution,** a period of great change in farming that began in the United Kingdom in the early 1700's. The development of new farming methods and equipment meant that fewer people were needed to produce food.

Seed drills are still used by farmers today. Modern seed drills (also called drills or seeders) are pulled by tractors instead of horses. Like Tull's seed drill, modern seed drills plant seeds in narrow rows spaced evenly apart. In addition to planting seeds, they can also be used to spread **fertilizer.**

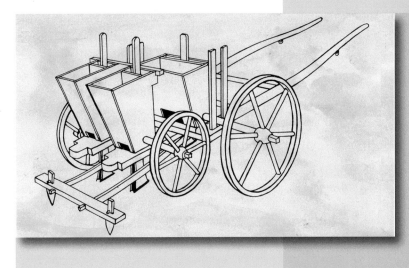

Jethro Tull's seed drill was the first practical farm tool for planting seeds in rows.

This modern seeder plants 12 rows of corn at one time.

▶ The Norfolk System

This 1808 painting shows the gentlemen farmers of Norfolk inspecting their sheep.

In the early 1700's, the county of Norfolk, England, was home to a group of inventors dedicated to the advancement of farming methods and technology. The group treated agriculture as a science and invented several improvements in the raising of crops and **livestock.** The improvements became known as the Norfolk System. The defining accomplishment of this system is the four-field **crop rotation** system.

In the early 1700's, the retired English politician Charles Townshend began to experiment with crop rotation in his Norfolk fields. He found that he could use turnips as one of the crops in a four-field rotation system. For the other three crops, he planted two grains (wheat and barley) and a **legume,** such as alfalfa or clover. Each crop affected

Year 1 rotation

the soil in different ways, absorbing and releasing different **nutrients.**

The four-field rotation system greatly increased the productivity of farmland. Unlike other systems, the four-field rotation system did not require farmers to leave any of their land **fallow** (unplanted) and unproductive. For the first time, farmers could plant crops on all their land without using up the soil's nutrients.

Townshend earned the nickname "Turnip" Townshend for his work. However, like Jethro Tull and his **seed drill,** Townshend's findings were not widely known until after his death in 1767. Then, in the late 1700's, an English nobleman named

Thomas Coke produced greatly increased yields using Townshend's system. He encouraged other farmers to adopt the method, and it soon became widely used in England.

This diagram shows an example of crop rotation. Before the invention of crop rotation, farmers had to leave part of their fields unplanted each year.

Clover	Wheat
Wheat	Turnip

Year 2 rotation

Charles Townshend

Charles Townshend (1725-1767) was a British politician. He also owned much land and invented the four-field rotation system in the 1700's. As a politician, Townshend played a role in American history. He was responsible for taxing the American colonies, one of the grievances listed by the colonists in the Declaration of Independence in 1776.

▶ The Cotton Gin

The cotton gin's teeth pulled seeds from cotton 50 times faster than a person could by hand.

Cotton is a plant fiber that can be woven into fabric. It has been an important farm crop for thousands of years. As early as 3000 B.C., cotton thread and fabric were made in what is now Pakistan and western India. Later, cotton growing and weaving spread to many parts of the world.

Historically, one of the most time-consuming steps in preparing cotton for use was removing the seeds from the cotton fibers. A simple machine for removing cotton seeds, called a **cotton gin,** was developed in India in the A.D. 400's. A later version of this machine was called the roller gin, which was used in the American Colonies by around 1740.

In a roller gin, a pair of wooden rollers press the seeds from the cotton. The roller gin could remove the seeds from long-staple cotton, a cotton **variety** that has long fibers. However, most American cotton farms grew short-staple cotton. The roller gin did not work well with this variety of cotton. Removing the seeds from a pound of cotton by hand took one person a full day. This limited the amount of cotton that farmers could grow.

In 1793, an American named Eli Whitney invented a more effective cotton gin. This gin had a crank that turned a cylinder covered with rows of wire teeth. When the cylinder turned, the teeth pulled the cotton through slots too narrow for seeds to enter. A

Eli Whitney

Eli Whitney (1765-1825) was an American inventor. He showed his mechanical talents as a boy when, at age 12, he built a violin. As a teenager, he established his own nail-making business.

After Whitney graduated from college, he moved to Georgia, where he built and **patented** his cotton gin. He made little money from his invention because competitors stole his ideas and made their own cotton gins. However, Whitney later made his fortune from selling firearms with interchangeable parts. This invention helped the North defeat the South in the American Civil War (1861-1865).

roller with brushes then removed the cotton fibers from the metal teeth.

Whitney's cotton gin allowed farmers to separate the seeds from short-staple cotton fibers more quickly and inexpensively. It could process as much cotton in a day as could 50 people working by hand.

The invention of the cotton gin made cotton a very valuable crop in the American South. The cotton **industry** expanded, and the United States became one of the world's leading producers of cotton. These developments, however, increased the South's dependence on **slaves** for labor (work). Eventually, the North and South fought in a civil war over slavery, among other issues.

Modern cotton gins, like these at a California mill, follow a complicated process that dries, strips, and presses the cotton into bales.

▶ The Steel Plow

In the mid-1800's, many farmers moved west to farm the fertile Midwestern prairies.

wood **plow** to prepare soil for planting. In the mid-1800's, many farmers from the eastern United States moved to the Midwest to begin farming the large, open prairies.

These farmers quickly discovered that the iron and wood plows they used on the sandy soils of the East Coast did not work well on the rich, black earth of the Midwestern prairies. A farmer either broke his plow, or he had to stop every few feet and scrape off the mud. Midwestern farmers needed something stronger and smoother to work with the thick, heavy soil.

By the mid-1800's, the **Agricultural Revolution** had spread throughout much of Europe and North America. During this time, improvements to farming tools in the United States brought swift changes to farming there.

Farmers in the eastern United States had long relied on the iron or

At around this time, an American blacksmith named John Deere moved to the Midwest and learned of the problems farmers were having with their plows. In 1837, he took a broken steel saw blade and used it to make the first steel plow. Steel is a strong, flexible material. Deere polished the steel on his plow to create a smooth surface that could cut through the heavy grasses and sticky soil without clogging.

The steel plow became instantly

Deere's steel plow worked well on the heavy soil of the American Midwest.

John Deere

John Deere (1804-1886) was an American inventor and manufacturer. Deere became a blacksmith's apprentice at the age of 17. In 1836, he opened a blacksmith shop in Grand Detour, Illinois. In 1837, he invented the first steel plow. He eventually became one of the world's greatest plow makers.

Eventually, Deere started a new company in Moline, Illinois. At his new company, Deere pioneered the use of a special type of hard steel to make plows.

Modern steel plows come in a variety of types and are pulled by tractors.

popular. Within 10 years, Deere was selling more than 1,000 steel plows each year. By 1855, he was selling 13,000 per year.

Deere eventually moved his business to Illinois. At first, he had to import steel from England. But as the country grew up around him and the steel **mills** in Pittsburgh, Pennsylvania, began making high-quality steel, his orders simply floated down the Ohio River and up the Mississippi River to northern Illinois.

Today, farmers in many countries use steel plows that are pulled by tractors. Companies that sell farm equipment make many different types of plows suited for specific types of soil.

▶ The Reaper

As farming became more productive, farmers needed better ways to **harvest** crops. For centuries, farmers harvested grains using handheld cutting tools, such as the sickle (a cutting tool with a short, curved blade) and cradle scythe (a cutting tool with a long, curved blade).

During the late 1700's and early 1800's, inventors began developing machines that could harvest grain mechanically. In 1834, the American inventor Cyrus Hall McCormick **patented** the first successful harvesting machine, called the **reaper.**

McCormick's reaper was a horse-drawn machine that cut grain as it was pulled across fields. The reaper had a straight blade linked by gears to a central wheel. As the wheel turned, the blade moved back and forth and sawed through stalks of grain. The stalks fell onto a platform, where a worker raked them onto the ground.

McCormick began selling his reaper in 1840. It needed 8 to 10 workers to bring in the crop. One worker drove the horse, another worker raked stalks from the platform, and the remaining workers tied the grain stalks into bundles.

Inventors continued to develop reapers that required fewer workers to operate. During the mid-1850's, self-rake reapers came into use. These reapers had a rake that swept stalks from the platform, eliminating the need for one of the workers.

The American inventor Sylvanus D. Locke invented another improved

Agricultural inventions and machines became big business in the 1800's. This reaper magazine advertisement is from 1875.

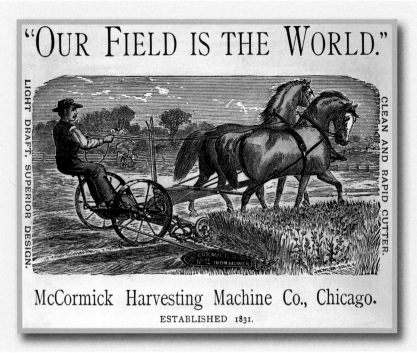

"OUR FIELD IS THE WORLD."

LIGHT DRAFT. SUPERIOR DESIGN.

CLEAN AND RAPID CUTTER.

McCormick Harvesting Machine Co., Chicago.

ESTABLISHED 1831.

Cyrus Hall McCormick

Cyrus Hall McCormick (1809-1884) was an American inventor who revolutionized grain harvesting in the United States with his invention of the reaper. He was born on a farm in Rockbridge County, Virginia. For years, his father tried unsuccessfully to build a mechanical reaper before passing the project on to his son, who got it to work. McCormick moved with his brother to Chicago, Illinois, where they opened a factory that built farm machinery. McCormick remained president of the McCormick Harvesting Machine Company until his death in 1884.

reaper in the early 1870's, called a binder. It gathered stalks into bundles before dropping them to the ground.

Still more improvements pushed grain production even further. The American brothers Hiram and John Pitts patented a combined threshing machine, which was used to separate kernels of grain from the stalks. (The word *thresh* means "to separate grains or seeds from a plant.") In 1834, the Pitts brothers patented a fanning mill, which is used to clean small grains, such as oats.

Other inventors combined a harvester and thresher, creating the combine harvester. Today, combines come with several different types of cutters on the front to handle different types of crops.

Reapers and other harvesting machines greatly increased the amount of grain that farmers could produce, harvest, and process. This led to the growth and success of farms across the United States.

Modern combines harvest wheat quickly, as seen in this picture from an English farm.

The Erie Canal connected the Great Lakes system to the Atlantic Ocean.

Steamboats could pull right up to Dart's grain elevator to load or unload grain.

By the 1800's, numerous inventions had made farmwork easier and more efficient, but transporting grain was still difficult, time-consuming work. Grain was often carried in bags or pushed in wheelbarrows or carts.

In the United States, the need for an efficient way to transport grain intensified with the opening of the **Erie Canal** in 1825. The waterway provided a route over which agricultural products could flow to the east. It stretched from the city of Buffalo across New York state, eventually connecting to the Hudson River.

After the opening of the Erie Canal, more grain moved through Buffalo than anyone could handle. Corn and wheat flooded into the city from the Midwest and needed to be stored for transport elsewhere in the world.

In 1842, an American inventor named Joseph Dart built a large wooden tower to store the huge amounts of grain coming through Buffalo. The structure also worked to load and unload the grain to and from the ships themselves. It was the world's first **grain elevator.**

Built right on the waterfront, Dart's grain elevator had buckets attached to a steam-driven belt. The belt would be lowered into the riverboat, where the buckets scooped up grain and brought it back to the elevator. From there, the grain was dumped into tall bins. The conveyor belt actually elevated, or

Preparing grain for storage involves many steps. Grain brought from fields is dumped on a conveyor, which carries the grain to an elevator. The elevator lifts the grain and releases it through a flow pipe into the drying unit. Here, hot forced air dries the grain. The dried grain is moved by a conveyor to a holding bin. The grain funnels from the bin onto the elevator, which lifts it to a flow pipe at the top of the elevator. The grain then flows into the storage bin.

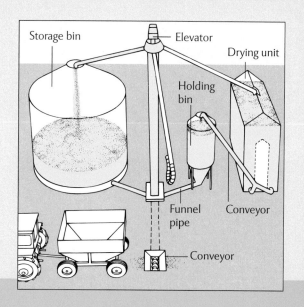

Storage bin — Elevator — Drying unit — Holding bin — Funnel pipe — Conveyor — Conveyor

raised, the grain from the ship right into storage.

Suddenly, 1,000 **bushels** (more than 35 cubic meters) of grain could be unloaded every hour. Before long, Dart's grain elevators were moving 10 times that much. The city of Buffalo quickly became the world's largest grain **port.**

Today, there are two types of grain elevators: country elevators and terminal elevators. Country elevators are found in nearly every town in grain-producing areas. These smaller elevators receive grain from farmers and usually store it for only short periods. The grain is cleaned in country elevators, and then it is loaded in railroad cars and shipped to large markets.

Terminal elevators stand at large grain markets and shipping centers like Chicago and New Orleans. In these elevators, grain is stored for the use of **millers** or to await shipment on the Great Lakes or overseas.

A huge seagoing ship fills with grain from a giant terminal elevator in Louisiana.

▶ The Tractor

From ancient times right up to the mid-1800's, farm equipment of all kinds was pulled by oxen, horses, donkeys, or mules. Much of this farmwork was difficult, even with a strong team of oxen. However, advancements in **industry** led to the development of one of the most important pieces of farm machinery: the **tractor.**

Shortly after the **Agricultural Revolution** began in the early 1700's, another great change in society took place. During the late 1700's and early 1800's, the **Industrial Revolution** spread through Europe and North America, changing the lives and work of the people. One of the advances that sparked the Industrial Revolution was the development of power-driven machinery.

By the early 1800's, an English inventor named Richard Trevithick (*TREHV uh thihk*) had designed and

A farmer drives a heavy steam tractor in the new state of South Dakota in the 1890's.

built the first high-pressure **steam engine.** This technology was first applied to **locomotives** and later to automobiles, though steam-powered cars were not widely used.

In 1870's, the first steam-powered tractors appeared. They were strong enough to pull 40 **plows** at once, but they hardly moved. In fact, most did not move at all. They were simply traction (pulling) engines. They could pull objects, but they could not push them.

By the late 1800's, **engineers** had developed the gasoline engine for use in automobiles. In 1892, an Iowa farmer named John Froelich used this technology to build the first practical gasoline-powered tractor. Unlike earlier tractors, Froelich's tractor could move forward and backward. Froelich attached his tractor to a threshing machine for separating kernels of grains from stalks.

Tractors ran on steel wheels until 1932, when air-filled rubber tires replaced them. These tires made tractors easier to maneuver and more comfortable to ride.

Though many improvements have been made to the tractor, its design remains largely the same. Tractors contain an engine, two small front wheels, two large rear wheels, and a seat in between the rear wheels. The back of a tractor includes a metal strip to

which equipment is attached.

Early manufacturing companies usually made only one tractor model or size. Today, companies make many different types of tractors. Besides farmwork, tractors are used in the military, on construction sites, and for logging, roadwork, and snow removal, among other things.

Tractors stand ready for sale near a cornfield in Florida.

F U N F A C T

Engine strength is measured in horsepower. Scottish inventor James Watt (1736-1819) created the term for his steam engines, describing their strength by how many horses it would take to provide equal power. Some modern tractors have engines of more than 500 horsepower, meaning they can do the work of 500 horses working together!

Fertilizers

Since nearly the beginning of agriculture, farmers have added substances to the soil to help plants grow. At first, they spread wood ash and animal manure in their fields as **fertilizer.** They knew that these substances helped plants grow larger and more rapidly, but they did not know how or why.

It wasn't until the late 1700's that scientists identified many of the chemical elements needed for plant growth. They discovered that crops need nitrogen, phosphorus, and potassium to grow. Manufacturers began making fertilizers containing these essential elements. At first, these fertilizers were expensive, so they were not widely used.

In 1909, the German scientist Fritz Haber developed a way to make ammonia, a colorless gas made of nitrogen and hydrogen that can be used as a fertilizer. Since nitrogen is critical for plant growth, the development of low-cost nitrogen greatly increased the use of fertilizers. More than 80 percent of all

Many farmers use animal manure, a natural fertilizer, on their fields.

ammonia made today is used to make fertilizers.

Chemical fertilizers gradually replaced **crop rotation** as a way to make soil more productive. Now farmers did not have to switch the types of crops they planted in a field or leave a field **fallow** every other year. With the help of fertilizers, they could plant their best-selling crops in every field year after year.

Today, many farmers around the world rely heavily on chemical fertilizers to increase crop production. Gardeners use fertilizers to grow better flowers and vegetables. Homeowners use fertilizer to make grass thicker and greener.

With chemical fertilizers, less land and work is needed to grow more crops. However, overuse of chemical fertilizers damages the environment. Rainwater can wash fertilizers into nearby water sources, where they can cause **pollution.**

Some farmers have returned to more traditional methods of agriculture. **Organic farmers** avoid the use of any human-made chemicals on their fields. Instead of using chemical fertilizer, they spread decayed plant or animal matter on their fields, which acts as a natural fertilizer.

Organic farmers may also use crop rotation to keep soil healthy. For example, they may plant **legumes** in a field that used to have corn plants. While corn removes nitrogen from the soil, legumes take nitrogen from the air and put it back into the soil.

Airplanes can "dust" entire fields with fertilizer in seconds.

Hybrid Crops

Nearly all of the 90 million acres (more than 36 million hectares) of cornfields in the United States are planted with hybrid corn.

For thousands of years, farmers have bred better **varieties** of crops by choosing only the best individual plants to parent the next generation. After the early 1800's, people gained a better understanding of how organisms inherited characteristics. As a result, a new type of breeding developed, called **crossing.**

Crossing refers to the breeding of two different individuals. When farmers cross two varieties or **species** of plants, they can design a plant with a specific combination of qualities. These "combination" plants made

from two different species or varieties are called **hybrids.**

To develop a hybrid plant, breeders choose two varieties of a plant. Each variety has a specific trait (characteristic) that they wish to pass along to future generations of plants. For example, one plant may have a trait that makes it more resistant to disease, while the other may produce larger plants. By crossing the two plant varieties, breeders obtain hybrid seeds with characteristics from both parent plants.

The parent plants of hybrids must be fairly similar. For example, a blackberry and a corn plant cannot be combined. However, a blackberry and a raspberry, which are related fruits, will cross to create a hybrid loganberry.

Hybrids offer many advantages to farmers. They combine such desired traits as increased productivity, resistance to disease, and the ability to grow at colder temperatures. Also, hybrids often produce more crops than the original parent plants.

Over time, natural hybrid versions

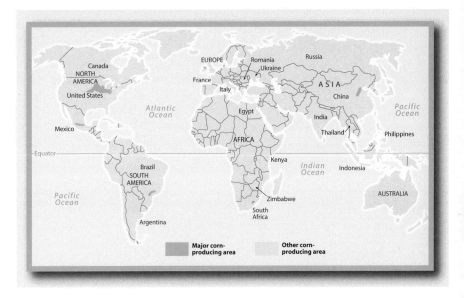

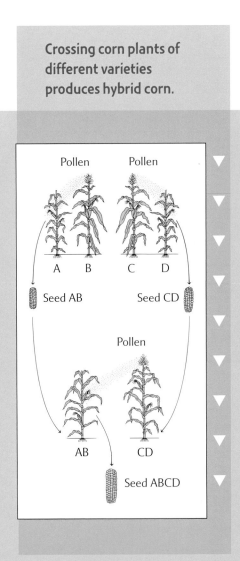

Crossing corn plants of different varieties produces hybrid corn.

of flowers, berries, and other plants (as well as animals) were found to exist. Plant breeders in the United States began experimenting with their own hybrids in the early 1900's.

By 1922, hybrid corn was available to U.S. farmers. By 1945, almost all the cornfields in the United States were planted with hybrids. Soon, hybrid breeding techniques extended to such crops as bananas, coffee, peanuts, wheat, and alfalfa. By the end of the 1900's, most vegetables and many kinds of **livestock** were hybrids as well.

Due to the growing of hybrid crops, U.S. corn **yields** nearly doubled from 1920 to 1960, and they continue to increase. Today, many grains, vegetables, fruits, flowers, and ornamental plants are hybrids. Hybrid plants are among the world's most important agricultural products.

Corn is grown in many places around the world, but originally it only grew in Central America.

▶ Pesticides

This farmer wears a protective suit as he sprays grape vines with pesticide in California.

Every year, farmers face **drought,** flooding, frost, and other forms of extreme weather that can severely damage their crops. On top of these threats, they must battle against pests like bugs, **bacteria,** and weeds.

Pests are organisms that can kill plants, interfere with plant growth, carry disease, or cause other trouble for farmers. There are many types of pests, and they all damage crops in different ways. For example, weeds compete with crops for sun, water, and **nutrients.** Insects can eat entire fields of crops and wipe out food production throughout an entire region. Even certain **fungi** and bacteria can harm plants.

For many years, there was little farmers could do against most pests. But in the late 1700's, scientists began to produce chemicals to fight pests. These chemicals are called **pesticides.**

Pesticides did not come into widespread use until the mid-1900's. As scientists developed pesticides, they created formulas that targeted specific pests, such as insects or rodents.

In 1939, scientists in Switzerland developed one of the first insecticides (pesticides made specifically to kill insects), called DDT. By the early 1950's, farmers used large amounts

A CLOSER LOOK

After World War I (1914-1918), many U.S. Army airmen and airplanes were out of work. In 1921, a pilot named John Macready filled his plane with insecticide and released the chemicals over an orchard infested with Catalpa sphinx moths. The moths were wiped out, and a new profession was born: crop-dusting. Crop dusters fly their planes low over crops, releasing pesticide as they go along. It is dangerous work, but it continues to this day.

of DDT to protect their crops.

Pesticides greatly reduce the amount of crops that farmers lose to unwanted pests. However, they also cause environmental damage. For example, pesticides can kill other organisms that benefit the environment. They may also drain into nearby water sources and cause **pollution.**

One of the most famous examples of the dangers of pesticides occurred in the late 1940's and 1950's. During this time, populations of peregrine falcons in the United States rapidly declined. Studies showed that the pesticide DDT had caused the birds' eggshells to become dangerously thin. This caused the baby birds to die before hatching. By 1960, peregrine falcons had almost completely vanished from North America.

In the 1970's, DDT was banned in the United States and strict pesticide laws were passed to ensure the safety of plants, animals, and people alike.

Today, some farmers use more natural methods of pest control. For example, they may pick bugs off plants by hand or plant flowers that repel (drive away) certain pests. However, chemical pesticides are still widely used throughout the world.

The airplane above dusts sheep with a special pesticide to kill ticks in 1948.

► Genetically Modified Food

Many foods sold in grocery stores today have been genetically modified.

In 1953, scientists discovered **DNA,** a chainlike molecule (group of cells) found in every living thing. DNA is like a recipe that determines what traits (characteristics) a living thing will have. For example, it determines whether someone will be tall or short and what their eye and hair color will be. By the 1970's, scientists had figured out how to change the in-structions in DNA in order to change the traits of certain organisms.

DNA is made up of **genes,** or individual parts of cells that carry information about a specific trait, such as eye color in people or leaf shape in plants. Different genes are responsible for different traits. In plants, these traits may include an ability to survive in dry conditions or to fight off harmful insects. Scientists can change these individual genes to influence the next generation. This process is called **genetic engineering.**

In a way, breeders have been genetic engineers for thousands of years. They select certain plants or animals to breed for the next season because they wish to carry their traits into the next generation. However, genetic engineering takes this process a step further. Genetic engineers make changes directly to an organism's genetic information.

During the 1980's, scientists developed ways to add certain types of genes to organisms. A **bacterial** gene was successfully inserted into tomato plants in 1987. It made that **variety** of tomato plant resistant to caterpillars—and created the first **genetically modified food (GM food).**

The first widely sold GM foods appeared in the mid-1990's. Some GM foods had an increased resistance to insects and disease. Others had an increased tolerance for certain types of **pesticides.** Today, scientists are working to create genetically modified crops that can be raised specifically to produce **vaccines** against infectious diseases in human beings.

Some farmers quickly accepted GM crops, but others have had concerns about their safety. Most independent scientific organizations have concluded that GM foods are safe to grow and eat.

Since GM foods are fairly new, some people worry that experts do not have enough information to determine if the foods are safe. Others worry that traits engineered into GM crops, such as resistance to pesticides, might spread to other plants. Genetic engineering presents clear benefits to the world today, but it also presents numerous challenges to governments, farmers, and consumers.

A CLOSER LOOK

By the late 1990's, scientists had found ways to alter the genetic information of animals. In 1996, scientists in England created a lamb, named Dolly, from the DNA of an adult sheep. This process is called cloning, which means the creation of an exact copy of a living thing. Before Dolly, scientists had cloned insects, frogs, fish, and mice. Since then, horses, cows, and monkeys have been successfully cloned.

Important Dates in Agriculture

c. 10,000 B.C. The domestication of plants and animals began in the Fertile Crescent.

c. 8000 B.C. Farmers first used the plow.

c. 6000 B.C. People began to herd cattle and grow grain in northern Africa.

c. 3000 B.C. Farmers in Mesopotamia and Egypt used the ox-drawn plow and large-scale irrigation for their crops. Agriculture spread to southern Africa.

c. 1500 B.C. Indians grew corn and beans in the Mexico area.

c. 1000 B.C. Indians raised gourds and sunflowers in North America.

c. 500 B.C.-A.D. 200 Roman farmers used crop rotation. Advanced farming techniques spread throughout the Roman Empire.

A.D. 100's The water wheel was used in China.

A.D. 600's The windmill originated in Persia (Iran).

A.D. 900's A new horse harness was introduced in Europe for plowing.

1100's Windmills spread to Europe.

1700's The Agricultural Revolution began. Jethro Tull invented the seed drill. Charles Townshend invented the Norfolk System.

1794 Eli Whitney invented his cotton gin.

1800's Windmills were used in the United States to pump water. Scientists discovered which chemical elements plants need for growth.

1834 Cyrus Hall McCormick invented the reaper.

1837 John Deere invented the steel plow.

1909 Franz Haber invented a way to capture atmospheric nitrogen. His process was an important advancement for fertilizer production.

1922 Hybrid corn was first sold in the United States.

1939 Swiss scientists invented DDT.

1970's Scientists figured out how to alter genes.

2006 About 90 percent of corn and 60 percent of soybean crops grown in the United States were from genetically altered seeds.

Glossary

Agricultural Revolution a period during the early 1700's when a series of agricultural discoveries and inventions made farming more productive.

aqueduct an artificial channel through which water is conducted to the place where it is used.

bacteria; bacterial single-celled organisms that can only be seen using a microscope; of or having to do with bacteria.

bushel a measure used for grain, fruit, vegetables, and other dry things.

civilization nations and peoples that have reached advanced stages in social development.

cotton gin a machine for separating fibers of cotton from the seeds.

crop rotation a system of planting an area of land with a succession of different crops.

crossing a type of breeding where two individual plants or animals are selected to produce the best offspring possible.

cultivate to loosen the ground to kill weeds and help plants to grow.

DNA the substance of which most genes are made. DNA is responsible for the passing down of inherited characteristics.

domestication the growing of plants from seeds or the raising of animals in captivity.

drought a long period of dry weather.

embankment a raised bank of earth or stones used to hold back water or support a roadway.

embryo an undeveloped plant within a seed.

engineer; engineering a person who plans and builds engines, machines, roads, bridges, canals, forts, or the like; the use of science to design structures, machines, and products.

Erie Canal a U.S. waterway built in 1825 that provided a route over which agricultural products and other goods could flow between the Great Lakes system and the Atlantic Ocean.

erosion the wearing away of a substance by wind or rain.

fallow left unplanted.

fertile producing crops easily.

fertilizer a substance that can be added to soil to help plants grow.

flood plain an area of flat land bordering a river and made of soil deposited by floods.

fungus; fungi any of a group of organisms that produce spores and get nourishment from dead or living organic matter; a plural of fungus.

gene a part of a cell that determines which characteristics living things inherit from their parents.

genetic engineering a group of techniques that are used to change the genes in an organism.

genetically modified food (GM food) food made from organisms—usually crop plants—that have been altered through genetic engineering.

germinate to begin to grow or develop; sprout.

grain elevator a building for storing grain, often with machinery for loading and unloading, cleaning, and mixing the grain.

harvest (n.) a reaping (cutting) and gathering in of grain and other food crops; (v.) to cut and gather grain and other food crops.

hybrid the offspring of two animals or plants of different varieties or species.

Industrial Revolution a period in the late 1700's and early 1800's when the development of industries brought great change to many parts of the world.

industry any branch of business, trade, or manufacture.

irrigation providing water to an area by artificial means.

legume a plant which bears pods containing a number of seeds.

livestock domestic animals used to produce food and other useful products.

locomotive a machine that moves trains on railroad tracks. It is sometimes called a railroad engine.

mill; miller a machine that grinds grain, such as wheat, between large grinding stones; a person who owns or runs a mill.

millet a very small grain used for food in Europe, Asia, and Africa.

nutrient a nourishing substance.

organic farmer a farmer who avoids the use of human-made chemicals on his or her farm.

patent (n.) a government-issued document that grants an inventor exclusive rights to an invention for a limited time; (v.) to get a patent for.

pesticide a chemical used to kill bacteria, insects, or animals that damage crops.

plow a tool used to turn over the soil, preparing it for planting.

pollution harm to the natural environment caused by human activity. Pollution dirties the land, air, or water.

port a city or town by a harbor.

reaper a machine that cuts grain or gathers a crop.

Roman of or having to do with ancient Rome or its people. The Roman Empire controlled most of Europe and the Middle East from 27 B.C. to A.D. 476.

row cultivation the planting of crops in rows.

seed drill a machine that makes rows of small trenches in the soil and drops seeds in them.

slave a person who is the property of another.

species a group of animals or plants that have certain permanent characteristics in common and are able to breed with one another.

steam engine an engine that is operated by the energy of expanding steam.

surplus an amount over and above what is needed.

till to prepare land for planting.

topsoil the upper part of the soil.

tractor a machine that pulls or pushes a tool or a machine over land.

vaccine a substance that helps protect a person from catching a disease.

variety a plant or animal differing from those of the species to which it belongs in some minor but permanent or transmissible way.

water wheel a wheel turned with water and designed to drive machinery, such as that of a mill or pump.

water mill a mill whose machinery is run by water power.

wind turbine a machine that turns wind power into mechanical power.

windmill a machine that is operated by wind power that can be used to provide power to pump water, grind grain, or generate electric power.

yield the amount produced.

▶ Additional Resources

Books:

- *Ancient Agriculture* by Michael and Mary B. Woods (Runestone Press, 2000).

- *A Farm Through Time* by Angela Wilkes (Dorling Kindersley, 2001).

- *Food and Agriculture: How We Use the Land* by Louise Spilsbury (Raintree, 2006).

- *Great Inventions: The Illustrated Science Encyclopedia* by Peter Harrison, Chris Oxlade, and Stephen Bennington (Southwater Publishing, 2001).

- *Great Inventions of the 20th Century* by Peter Jedicke (Chelsea House Publications, 2007).

- *How to Enter and Win an Invention Contest* by Edwin J. Sobey (Enslow, 1999).

- *Inventions* by Valerie Wyatt (Kids Can Press, 2003).

- *Leonardo, Beautiful Dreamer* by Robert Byrd (Dutton, 2003).

- *So You Want to Be an Inventor?* by Judith St. George (Philomel Books, 2002).

- *What a Great Idea! Inventions that Changed the World* by Stephen M. Tomecek (Scholastic, 2003).

Web Sites:

- AIPL Kid's Corner
  http://www.aipl.arsusda.gov/kc/kcindex.html
  Information on dairy products and cows from the Animal Improvement Programs Laboratory in the United States.

- Biotechnology in Agriculture
  http://www.fao.org/ag/magazine/9901sp1.htm
  A spotlight article from the Food and Agriculture Organization of the United Nations.

- The Learning Center - Kids
  http://www.epa.gov/oecaagct/lkids.html
  The U.S. Environmental Protection Agency's Learning Center includes links to information about agriculture, food safety, and pesticides, among other topics.

- Natural Resources Conservation Service
  http://www.nrcs.usda.gov
  Homepage of the Natural Resources Conservation Service, a federal agency whose goal is to help people conserve and sustain soil and other natural resources.

- USDA Agricultural Research Service
  http://www.ars.usda.gov/is/kids
  Information for students on a variety of agricultural topics from the United States Department of Agriculture.

Index

A
Agricultural Revolution, 23, 28, 34
agriculture, 4-5
airplane, 37, 41
American Wind Power Center and Museum, 19
ammonia, 36-37
animal. See horse; livestock; ox
aqueduct, 13

B
Beebe windmill, 19
breeding, 36-37, 42

C
cloning, 43
combine harvester, 31
corn, 9, 14, 38-39
cotton gin, 26-27
crooked plow, 10-11
crop: domesticated, 6-7; genetically modified, 42-43; hybrid, 38-39
crop duster, 37, 41
crop rotation, 14-15, 24-25, 37
crossing, of plants, 38
cultivation, 8; row, 11

D
Dart, Joseph, 32-33
DDT, 40-41
Deere, John, 28-29
DNA, 42, 43
dog harness, 21
Dolly (lamb), 43
domestication, 6-7
drill, seed. See seed drill
Dutch windmill, 18

E
Egypt, ancient, 10, 12, 17
Erie Canal, 32
Euphrates River, 6

F
falcon, peregrine, 41
fanning mill, 31
Fertile Crescent, 6
fertilizer, 11, 15, 23, 36-37
flour, 16, 17
Froelich, John, 35

G
gasoline engine, 35
gene, 42
genetically modified food, 42-43
genetic engineering, 42-43
GM food. See genetically modified food
grain, 33, 39; grinding, 16-17; harvesting, 30-31; transporting, 32-33
grain elevator, 32-33

H
Haber, Fritz, 36
harness, 20-21
harvesting, 6, 11, 30-31
hoe, 8-9
hoecake, 9
horse, 22, 30, 34; harness for, 20-21
horsepower, 35
hybrid crop, 38-39

I
Industrial Revolution, 34
invention, 4
irrigation, 12-13, 17, 18

J
John Deere Company, 29, 35

L
legume, 15, 24, 37
livestock: domesticated, 6-7; genetically modified, 42, 43; hybrid, 39; in Norfolk System, 24
Locke, Sylvanus D, 30-31

M
Macready, John, 41
McCormick, Cyrus Hall, 30-31

N
Netherlands, 18
Nile River, 12
nitrogen, 14-15, 36, 37
Norfolk System, 24-25

O
organic farming, 37
ox, 10, 20, 34

P
pesticide, 9, 15, 40-41, 43
Pitts, Hiram and John, 31
plant. See crop
plow, 10-11; horse-drawn, 20-21; steel, 28-29
pollution, 15, 37, 41

R
reaper, 30-31
rice, 13
roller gin, 26
Rome, ancient, 13-15, 20
row cultivation, 11

S
scratch plow, 10
scuffle hoe, 8-9
seed, 7, 8, 22
seed drill, 22-23, 25
slavery, 27
steam engine, 35
steel plow, 28-29

T
threshing machine, 31
Tigris River, 6
tomato, 42, 43
Townshend, Charles, 24-25
tractor, 23, 34-35
Trevithick, Richard, 34-35
Tull, Jethro, 22-23, 25

W
water buffalo, 11
water mill, 16-17
water wheel, 16-17
Watt, James, 35
weed, 9, 40
Whitney, Eli, 26-27
windmill, 18-19
wind turbine, 19